Editor
Lorin Klistoff, M.A.

Managing Editor
Karen Goldfluss, M.S. Ed.

Illustrator
Renée Christine Yates

Cover Artist
Brenda DiAntonis

Art Manager
Kevin Barnes

Art Director
CJae Froshay

Imaging
Alfred Lau
Rosa C. See

Publisher
Mary D. Smith, M.S. Ed.

MATH IN

Graphs, Data & Chance

Grades 1-2

• Hands-on activities • Easy-to-use manipulatives • Lessons & objectives included

Author

Bev Dunbar

(Revised and rewritten by Teacher Created Resources, Inc.)

This edition published by *Teacher Created Resources, Inc.*
6421 Industry Way
Westminster, CA 92683
www.teachercreated.com

ISBN-1-4206-3533-6

©2005 Teacher Created Resources, Inc.

Made in U.S.A.

Teacher Created Resources

The classroom teacher may reproduce copies of materials in this book for classroom use only. The reproduction of any part for an entire school or school system is strictly prohibited. No part of this publication may be transmitted, stored, or recorded in any form without written permission from the publisher.

Table of Contents

#3533 Math in Action

©*Teacher Created Resources, Inc.*

Introduction

Math in Action: Graphs, Data and Chance is part of the *Math in Action* series of teacher resources for young students. The ability to interpret graphs and data is a vital aspect of being numerate in today's society, and these action-packed lesson plans will help students learn necessary skills in fun, practical ways. The flexible activities range from simple to challenging, to help support different ability groups.

Making your teaching life easier is a major aim of this series.

The book is divided into sequenced units, packed with reproducible activity resources, discussion starters, and worksheets for small groups to a whole class. The instructions are easy to follow and skills are clearly stated. You can see the complete range of skills on page 94.

The activities are organized so that each topic could be taught as a unit or as individual lessons woven into your daily curriculum. Graphing, in particular, can be included in almost any curriculum subject.

Each activity is designed to maximize the way in which your students construct their own understandings about graphs, data, and chance. Activities are open-ended and encourage students to think and work mathematically. The emphasis is on practical manipulation of materials and the development of language and recording skills.

Enjoy exploring these concepts with your students!

©*Teacher Created Resources, Inc.*

How to Use This Book

❑ **Teaching/Learning Plans**

Graphing: The first three units of this book contain many ideas for developing graph concepts. You can choose the unit which best fits your students needs and abilities. Link the concepts to your current curriculum in any topic, and use some activities as assessment tools.

Chance and Data: The activities in these three units are open-ended and encourage your students to work mathematically and think critically. These lessons can be used with small groups or your entire class.

❑ **Reproducibles**

There are three types of reproducibles.

1. *Activity Resources*
   (e.g., page 77, *Dinosaur Danger Spinners*)
   These spinners, labels, playing cards, and game boards support free exploration, as well as structured activities. Photocopy multiple sets on different-colored paper to enhance class management.

2. *Discussion Starters*
   (e.g., page 28, *Tally Starters*)
   These are an extra stimulus for group work. Encourage your students to create their own. Laminate the cards for years of reuse.

3. *Reusable Worksheets*
   (e.g., page 62, *Cookies*)
   Unlike traditional worksheets, these allow for varying solutions and can be used by the same students again and again with different results. The structured tasks support your learning outcomes.

❑ **Skills Record Sheet**

Each activity has clearly coded learning skills. See page 94 for a list of skills addressed in this book, and use this checklist to record individual progress through each topic. Assess a few students each day.

❑ **Sample Weekly Program**

On page 95, you will find one example of how to organize a selection of activities as a five-day Chance and Data unit for a class. On page 96, you will find a blank weekly program for your own use.

Exploring Graphs with Real Objects or Pictures

In this unit, students will do the following:

- Pose questions and collect related information

- Sort and compare groups by matching objects in lines

- Group pictures and symbols to represent and compare data

- Describe and interpret information from object displays

(The skills in this section are listed on the Skills Record Sheet on page 94.)

Ways to Graph with Students

When first developing the concepts of a graph with young students, use their actual bodies rather than pictures or objects. Focus on where they should stand in each line, how many people there are, and what label you can give each group. Give your people graph a title, too.

Who's Tallest?

- *(Whole Class)* Ask your students, "Who do you think is the tallest person in our class?" Have students predict first, then check. Ask, "How can you find out?" Have students race to form themselves into height order from the shortest to the tallest student. Ask, "How should you stand?" (e.g., in a long line from left to right or behind each other from front to back) Have students find two things to tell the person next to them about their line of people. (e.g., Gabi is as tall as Phil. Isshan is shorter than Maria.)

Who's Taller than Sam?

- *(Whole Class)* Ask students, "How can you rearrange the class into three height teams?" (e.g., Select a medium height student as the sorting guide—shorter than Sam, the same height as Sam, taller than Sam.) Ask, "What will the groups look like?" (e.g., in lines from front to back or lying down head to toe) Ask students, "Which line is the longest? Which group has the most people? The fewest people?" Have students tell the person next to them two things they notice about the three groups. (e.g., More students in the class are shorter than Sam.)

The Longest Hair

- *(Whole Class)* Ask students, "Who do you think has the longest hair in our class?" Have students predict first, then check. Ask, "How can you find out?" (e.g., Use a length of string to check.)
- Have students put themselves into a long line—from the student with the shortest hair to the student with the longest hair. Ask, "How many students have hair the same length as yours? How should these students stand?" (e.g., in a line behind each other)

The Longest First Name

- *(Whole Class)* Ask students, "How many letters are in your first name? Who has the shortest name? Who has the longest? Who has the same number of letters as you do?" Have students make a human graph by standing in groups from the shortest to the longest name.

- Discuss the results together.

©*Teacher Created Resources, Inc.*

Ways to Graph with Students

Either/Or

- *(Whole Class)* Have students think of a way to sort the class into two lines. (e.g., Do you prefer to drink water or milk?) Have students predict which team will have the most members first, then check. (e.g., Students who prefer water stand in a line on the left. Students who prefer milk stand in a line on the right.) Ask, "What label can you give each team?" (e.g., We like milk. We like water.) Have students count how many students are in each group. Have them tell the person next to them something they notice about these two lines. (e.g., More students like milk.) Have students try some more either/or sorts. (e.g., We like purple/We like orange. We would rather visit the seaside/We would rather visit the mountains. We like dinosaurs/We don't like dinosaurs.) Ask, "What else can you discover about the class?"

Our Favorite

- *(Whole Class)* Have students talk about their favorite things. Ask, "Does anyone else like the same things as you?" Have students decide on a topic to investigate together. (e.g., our favorite books) Have students find all the students in the class who have the same favorite book. Have students stand in team lines and write a label for the person at the front of their team to hold up. (e.g., *The Cat in the Hat*) Ask, "How many different teams are there? What are the top five books in order of popularity?" Have students try some other favorite sorts. (e.g., our favorite movie, our favorite TV show, our favorite color)

My Family

- *(Whole Class)* Discuss with students how families come in all different shapes and sizes. Talk about the number of brothers or sisters they have. Talk about the number of people in their families altogether. Have students race to form groups of students with the same number of family members. Have them write a large label for the person at the front to hold up. Ask, "Which group has the most members? The fewest members? What does this tell you about families in the class?"

My Age

- *(Whole Class)* Have students talk about how old they are now and how old they will be at the end of the year. Have students form groups by their current age. Discuss the results together. Ask, "What do you notice?" Have them sort themselves again into groups by the age they will be at the end of the year. Ask, "Were there any differences?" Have students try sorting themselves into teams according to birthday month, too. Have them find three interesting things to tell the class about these sorts.

Ways to Graph with Objects

Once you think the students understand the idea of analyzing information by comparing the number of students in two or more lines, substitute objects to represent each student or event.

The Largest Shoe

- (*Whole Class*) Discuss with students how shoes come in all different sizes. Ask, "Who do you think has the largest shoe in our class? What does largest mean? Do you mean the longest shoe or the shoe which covers the greatest area?" Discuss with students and decide on what they will measure. Have them predict the largest size first, then check. Ask, "How can you find out?" (e.g., Students take off their left shoes and put them into groups by size.) Ask, "Are there many shoes exactly the same size as yours? What labels will you give each group?" (e.g., shoes the same size as Alex's shoe)

Falling Leaves

- (*Whole Class*) Have students walk around the playground or school environment until they find an area that has some fallen leaves. Have each student collect an interesting leaf and then put the leaves all together to make a collection. Have students think of some questions to ask. (e.g., "Are there more medium-length leaves?"—Have them sort into short, medium, long leaves. "Are there more brown leaves?"—Have them sort into colors.) Have students sort, count, compare, and discuss the results.

Toys

- (*Whole Class*) Ask everyone to bring in a favorite toy. Ask, "What sort of questions can you ask about this collection?" (e.g., "Which type of toy is the most popular? Is one color more common? From what type of materials are the toys made?" Have students decide on a question together and sort the toys to match. Have students arrange each group in a line and then write a label for each group. Discuss the results together. (e.g., Yellow is the most popular color for our toys.)

Yucky Vegetables

- (*Whole Class*) Most vegetables are delicious but there must be one or two that are not popular in your class. Ask everyone to bring in a vegetable that they really do not enjoy eating. What sort of questions can students ask about these? (e.g., "Which vegetable is the least liked?") Have students make a veggie graph by sorting the same types together. Ask, "What labels will you use? What is the title of your graph?"

Ways to Graph with Pictures

Once students have mastered using real objects to construct your graphs, you can draw, group, compare and line up pictures to represent each student, object, or event. Display these on large sheets of butcher paper.

The Largest Hand

- (*Whole Class*) Discuss with students how hands cover large and small areas. Ask, "Who do you think has the largest hand in our class?" Have students predict first, then check. Ask, "How can you find out?" (e.g., Trace each hand onto paper and cut it out.) Ask, "Should you have your fingers together or outstretched?" Have students arrange these hand cutouts in lines from the smallest to the largest hand. Ask, "Are there many hands exactly the same size as yours?" Have students glue the hands onto a large sheet of paper to make their hand graph and add labels. Ask, "What is the title of this graph?"

Eye Colors

- (*Whole Class*) Ask students, "What color eyes do you have? What color eyes are possible? What color do you think will be the most common in the class?" Have students sort themselves into groups by eye color. Have them stand in teams and discuss the results. Have students record their graph on a large sheet of paper by drawing, coloring, and cutting out an eye (see page 10), and then pasting these in lines to represent their groups. Have them label the graph and put it on display in the classroom. Have students write interesting facts they discovered. (e.g., There were no students with green eyes in our class.)

Our Pets

- (*Whole Class*) Ask students, "What is a pet? Why do humans have them? What type do you have? Why? Why not?" Have students draw a picture (or use page 11) to represent their pets. Ask students, "How can you group these drawings?" (e.g., dogs, cats, birds) Have students make a large class graph with labels and write two interesting facts they discovered about the pets.

The Family Vehicle

- (*Whole Class*) Ask students, "What sort of transportation does your family use?" (e.g., sedan, station wagon, four-wheel drive, sports car, truck, bus, bicycle, walking) Have students draw a picture of what type of vehicle they have or the transportation they use (or use page 12). Color, then cut out, and arrange to form a class graph. Ask, "What can you tell from the results of your graph?"

9

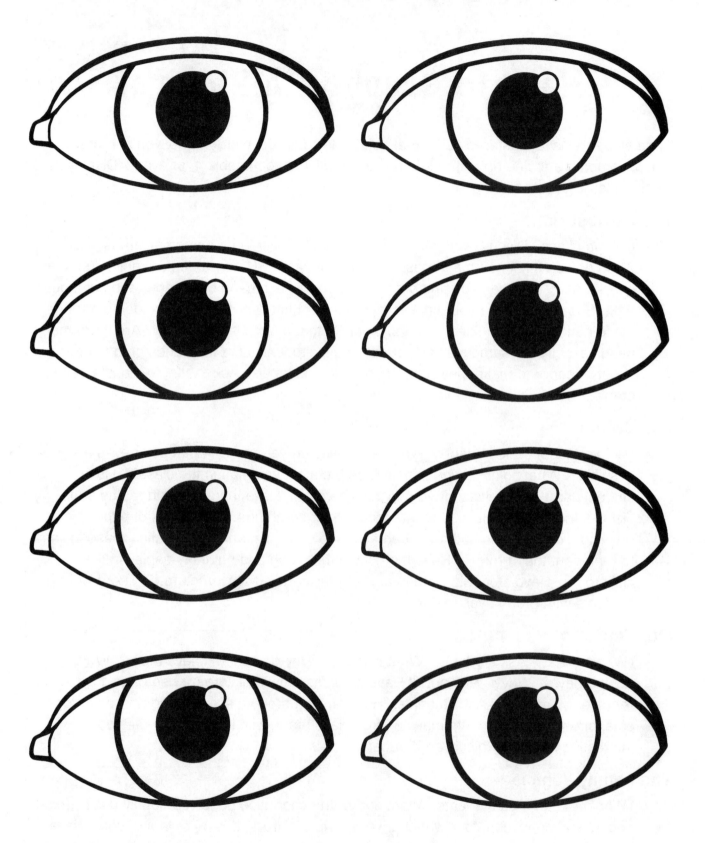

                    ©*Teacher Created Resources, Inc.*

©Teacher Created Resources, Inc.

©*Teacher Created Resources, Inc.*

Ways to Graph with Name Cards

Make a same size name card for each student (you can use page 14, enlarged). Use these to determine the results of class surveys. (e.g., Create your graphs by lining up name cards and counting how many names in each group.)

My Favorite Number

- *(Whole Class)* Have students think of all the ways numbers are used (e.g., house numbers, birthdays). Have students think of just the numbers from 1–10. Ask, "Do you have a favorite?" Have students predict first, then find a way to check. (e.g., Label 10 containers from 1–10.) Have them place their name cards in the containers that show their favorite numbers. Ask, "Were there any numbers that no one chose?"

That Family Car Again

- *(Whole Class)* Have students find out what brand of car or cars their families drive (e.g., Mitsubishi). Ask, "Do you use more than one car?" Have students draw or cut out a magazine picture of each type. Have them write labels to match and place their name cards with the matching car types. Ask, "What will you do if your family owns more than one car?" (e.g., Write another name card.) Ask, "What can you tell from the results?'

The Funniest Joke

- *(Whole Class)* Ask students if they know any jokes. Have them each find five jokes that their class thinks are very funny. Have students give each joke a title and write it on a label. Have them vote for their favorite jokes by placing their name cards beside them. Have students find an interesting way to display their results. Discuss together (e.g., The most popular joke in our class is . . . .) Ask, "Were you surprised by any of your findings?" Have students write a report to publish on their school website.

Cartoon Characters

- *(Whole Class)* Ask students, "Which cartoon character do you like the best?" Have students draw a picture label for each character. Then have them vote for their favorite by placing their name cards beside the matching drawings. Discuss the results together. (e.g., "The most popular cartoon character in our class is . . . ." "How many people voted for the same character as you?")

#3533 Math in Action

©*Teacher Created Resources, Inc.*

Ways to Graph with Symbolic Objects

Using symbols to represent a student, an object, or an event is another way to create a graph. Be creative! What else can you use as graphing materials in your classroom? (e.g., paper clips, beads on an abacus)

Windy Weather

- *(Whole Class)* Ask students, "What's the weather like today?" Discuss all the possible forecasts. (e.g., windy, wet, sunny, cloudy, stormy) Draw a label for each possible outcome. Each day place a stacking block to make a tower. (e.g., Put plastic building bricks or multilinks on the label to represent the matching weather. Some days may require more than one block if the weather changes dramatically.) At the end of each month, compare the weather towers. Ask students, "Which type of weather has been the most common?" Talk about what they discovered. (e.g., In March it was more often sunny than rainy.)

The Widest Smile

- *(Whole Class)* Tell students to smile at the person sitting next to them. Ask, "How wide do you think his or her smile is?" Have students find a way to measure the exact length of their friends' smiles from left to right. Ask, "Will you measure across in a straight line or try to measure the actual curved length of the smile?" Have students decide on what it is they are measuring and then find a way to check. (e.g., Cut a piece of string or ribbon to match.) Arrange the class smiles by length to make a graph. Then have students send an email to a class at another school explaining some interesting features of their results.

Our Most Popular Dinosaur

- *(Whole Class)* Have students think of all the dinosaurs they know. Have them predict which one will be the most popular and explain their reasons. Have them each draw a picture of their top five different dinosaurs (or adapt those on page 16), each on a piece of paper. Tell students to ask everyone in their grade to vote for which of these dinosaurs they like the best by placing a sticker on the matching drawing. Ask, "Which dinosaur got the most stickers? Does this match your prediction? Did any dinosaurs receive no votes?" Talk about the results. Have students record two or more statements about their findings in a workbook. Have them draw a picture to match.

©*Teacher Created Resources, Inc.*

Exploring Column Graphs

In this unit, students will do the following:

- Design and carry out a survey

- Place objects, pictures, and symbols in grids to represent data

- Use tally marks to collect data

- Construct, label, and interpret column graphs

(The skills in this section are listed on the Skills Record Sheet on page 94.)

Pat-a-Pet

Skills

- Design and carry out a survey
- Place objects, pictures, and symbols in grids to represent data

Grouping

- whole class
- individuals

Materials

- a large chalk grid drawn on the playground (e.g., each square is 2' x 2')
- sample Pat-a-Pet graph (page 19)
- workbooks
- pencils
- blank grid (page 19)

Directions

- Have students imagine the local pet shop is having a "pat-a-pet" promotion. Ask students, "Which do you prefer—patting a kitten or patting a puppy? Why?"
- Discuss the playground grid. Tell students to notice how the grid can be seen as lots of rows or lots of columns; columns are vertical, or up and down, and rows are horizontal, or left to right. Ask them, "How can you use this to help you find the most popular pet to pat?" (e.g., Make a human graph. Each person stand in one square of a column. Puppy-patters stand in the column on the left. Kitten-patters stand in the column on the right.) Have students predict which column will have the most squares filled.
- Ask, "What title will you give this graph? What label will you give each column?" Have students make their choices by each standing in one square. Have them count how many people are in each column. Ask, "What does this tell you?"
- Ask students, "What will happen if you stand in rows instead of columns? What label will you give each row?" Have students race to rearrange themselves into two groups, this time standing in rows. Have them count how many members are in each row. Ask, "Do the two graphs give the same information? Why? Why not?"
- Have students draw their own version of either graph. Ask, "How will you draw your grid?" Remind students to record the title and label each column.

Variations

- Have students look at the graph (page 19) made by students at another school. Ask, "What title could you give it? What column labels would you give?" Have students write at least three statements about information they discovered from this graph. (e.g., There are ___ students in this class.)
- Have students use the blank grid to create their own graphs. Ask, "What topic can you investigate?"

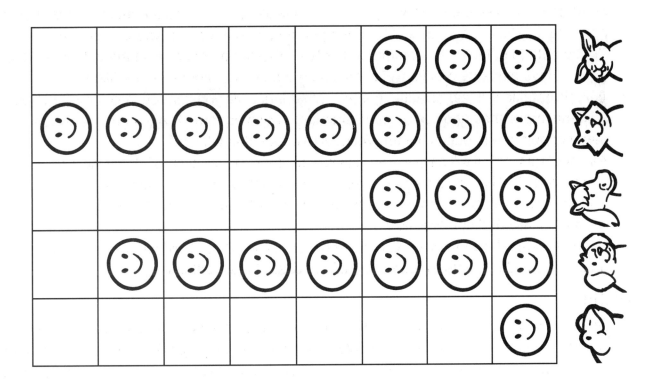

©*Teacher Created Resources, Inc.*   *#3533 Math in Action*

Fish Tally

Skills

- Design and carry out a survey
- Place objects, pictures, and symbols in grids to represent data
- Use tally marks to collect data
- Use grids to construct, label, and interpret column graphs

Grouping

- whole class
- pairs

Materials

- Fish Tally activity sheet (page 21)
- scrap paper
- colored pencils
- a suitable grid, hand drawn or photocopied (e.g., page 19 or page 24)

Directions

- Ask students, "Who keeps fish as pets in your class? What sort of fish do you keep?" Discuss together.

- Tell students baby fish sometimes grow very quickly. Have students look at the Fish Tally picture. Ask, "How many fish do you think there are?" Have students estimate first, then check. Then have them color these fish at random using five different colors or designs. (e.g., striped fish, spotted fish) Have students exchange pictures with a partner.

- Have students look at the fish in the bowl. Tell them that the fish have grown too big for their bowl! Tell students that the fish need to be sorted and put into separate tanks. Have students find all the fish that look the same. Have them make a tally of how many fish of each type there are. (e.g., How many blue fish are there?)

- Have students use a grid to record their tally results. (e.g., Color squares in one column to represent blue fish.) Have students write a title for their graph and labels for each column.

- Have students compare graphs with a partner. Discuss the findings together. Ask, "Which type of fish was the most common? The least common?

Variation

- Repeat using a different color/design combination.

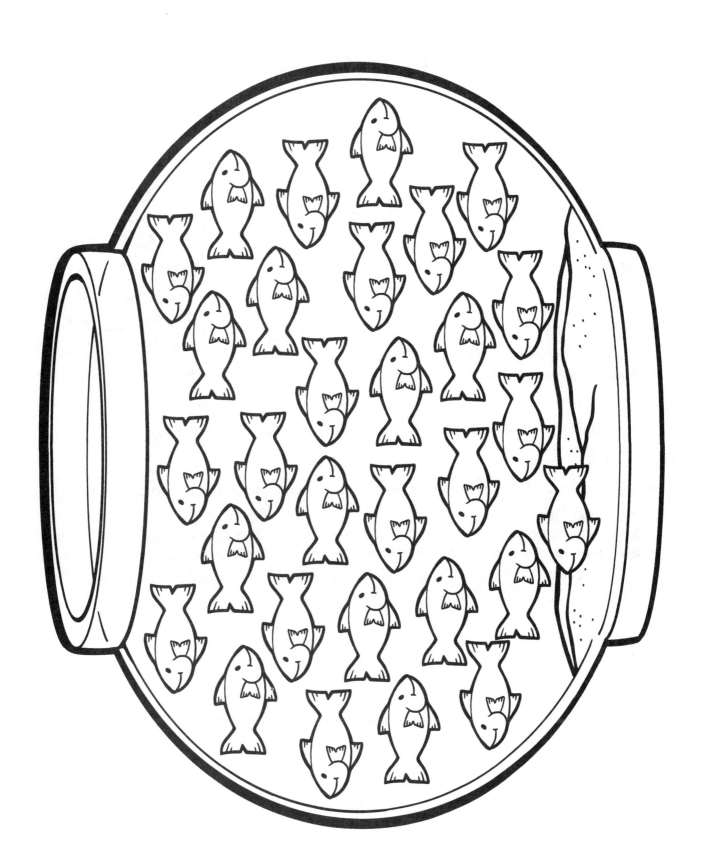

©*Teacher Created Resources, Inc.*

Minute Measures

Skills

- Design and carry out a survey
- Place objects, pictures, and symbols in grids to represent data
- Use tally marks to collect data
- Use grids to construct, label, and interpret column graphs

Grouping

- whole class
- pairs

Materials

- a minute timer (e.g., a stopwatch)
- Minute Measures Tally Sheet (page 23)
- Minute Measures Grid (page 24)
- paper and pencils
- counting cubes
- P.E. equipment (see Variation)

Directions

- Ask students, "How long is a minute?" Using a timer to check, have students close their eyes while a minute passes.
- Ask, "How many times do you think you can count by tens to one hundred in a minute?" Ask someone to count out loud in front of the class. Record how many times they reach one hundred as tally marks on the chalkboard. Count up the tally marks to get their total. Ask, "What can you do with this information?" (e.g., See if someone can increase this number.)
- Have students work with a partner to perform all five of these actions for one minute each: write your name, count by tens to one hundred, do jumping jacks, say the alphabet (clearly), and stack five counting cubes. They will record their tallies on the Minute Measures Tally Sheet (page 23).
- Have students record their Minute Measures tallies as a graph using the Minute Measures Grid (page 24). Ask, "What sort of grid you need? How many columns? How many rows? What labels to you need?"
- Ask, "What information does this tell you?" Have each pair of students create a short report to present to other students in the class.

Variation

- Ask students to create Minute Measure tests to perform outdoors (e.g., bounce a ball, toss and catch a beanbag, skip a rope, run to a tree and back). Have a small group of students perform each test and tally, count, and graph their results. Each group can present their findings to the class.

  ©*Teacher Created Resources, Inc.*

Minute Measures Tally Tasks

Tom Tom Tom ✏	10 20 30 40			A B C D E

Minute Measures Tally Tasks

Tom Tom Tom ✏	10 20 30 40			A B C D E

23

#3533 Math in Action   ©*Teacher Created Resources, Inc.*

Tally Starters

Skills

- Design and carry out a survey
- Use tally marks to collect data

Groupings

- whole class
- small groups

Materials

- pattern blocks
- Tally Starters activity cards (pages 26–28)
- paper
- pencils

Directions

- Ask students, "Which pattern block do you think is the most useful for building a pattern? How can you find out?" (e.g., carry out a survey)
- Ask everyone to make a design. Keep a tally of which blocks were used each time. Ask about five people to do this each day. At the end of a week you can make a graph based on the most popular block from each student's tally.
- Ask students, "Why use tally marks to record information? (e.g., It is easier than writing words. It is easy to count up by 5s when you are finished.)
- Ask students, "What sort of things can you tally?" (e.g., your score in a game, your preference in a survey)
- Ask students, "Why do you sometimes turn your results into a graph?" (e.g., A graph is easier to look at than just numbers or tally marks. Graphs help us compare and analyze events.)
- Ask students, "What sort of questions can you ask, or statements can you make, about your graph?" (e.g., The most popular block was . . . . The least popular was . . . It must be difficult to make a pattern using . . . .)
- Tell students that by carrying out a survey they can discover things that are not obvious in the first place. Tell them that they can conduct a survey about almost anything. Ask, "What do you want to find out? What question will you ask? Who should you ask? How many people should you ask? How will you collect and organize the information?"
- Have students work in teams. Have them design and carry out their own survey. Have students explain their question and results to another team.

Variation

- Have students design their own survey based on an idea from one of the Tally Starters cards.

25

Candies

Some of students' favorite candies are lollipops, chocolate, and gum.

What are your favorites?

I Wish

Imagine you could receive a special present.

For what would you wish?

What a Name

Linda and Pete were the most popular baby names about 50 years ago.

Who else has the same name as you?

Fruit

My favorite fruit is an apple.

What is yours?

Games

My grandma's favorite game when she was young was jumping rope.

What is your favorite game?

Vacations

For a one week vacation, would you prefer to live in a tent, a motor home, or a log cabin?

Yummy Ice Cream

Many people think chocolate is the most popular ice-cream flavor.

What is yours?

Lost Teeth

You do not keep your baby teeth forever.

How many teeth have you lost?

©*Teacher Created Resources, Inc.*

Hair

I have light brown hair.

What color hair do you have?

Cousins

I have five cousins—three on my dad's side and two on my mom's side.

How many cousins do you have?

TV

I like watching cartoons best.

What is your favorite TV show?

Wow

The best thing that ever happened to me was when my baby sister was born.

What is your best thing?

#3533 Math in Action

©*Teacher Created Resources, Inc.*

Exploring Grid Paper Column Graphs

In this unit, students will do the following:

- Use small grid paper to construct column graphs

- Analyze column graph data to form opinions

- Use different scales on column graphs

(The skills in this section are listed on the Skills Record Sheet on page 94.)

Measure Me

Skills

- Use small grid paper to construct column graphs
- Analyze column graph data to form opinions

Grouping

- whole class
- small groups

Materials

- yard measures (e.g., string cut to one-yard lengths)
- pound masses/balance pans
- gallon containers
- one-minute timers
- Measure Me work strips (page 31)
- colored pencils
- small grid paper (page 32 or 33)

Directions

- Review with students what they know about column graphs. (e.g., They have a title and labels to show what each column represents. They have a scale which shows how many spaces are in each column.)
- Discuss with students how grids make it easy to construct graphs. (e.g., Color one space for each tally mark.) Remind students that they should write the number scale on the left side of the grids. (e.g., 1 to 10)
- Tell students they need to make sure there are enough spaces on their grids to match their tally marks. (e.g., If there are 17 results in one group, they will need 17 spaces in their matching column.) Discuss how to select the appropriate grid paper if they have large numbers in their results.
- Divide students into four groups, each with a different measurement focus—length, mass, volume, and time—using a matching workcard and measuring equipment.
- In their groups, have students discuss what they will measure. (e.g., Length: What is the height of each student? Mass: How heavy is each item in the P.E. box? Volume: Which items in our collection of empty containers hold exactly 1 gallon? Time: Can you run to the fence and back in a minute?)
- Have students record their results on the work strips. Have them present a group report in the form of a graph. (e.g., Use a computer program or page 32 or 33.) Make sure each group has a title and labels. Have them analyze their graph and make statements. (e.g., Most of our P.E. equipment is heavier than one pound.)

Variation

- Have students use the Measure Me work strips to record information from around their homes. Have them collect the results together to make one large class graph. Have them compare this with previous information from the small groups.

Less than 1 yard	1 yard	More than 1 yard

Less than 1 pound	1 pound	More than 1 pound

Less than 1 gallon	1 gallon	More than 1 gallon

Less than 1 minute	1 minute	More than 1 minute

©*Teacher Created Resources, Inc.*
                                                               #3533 Math in Action

_____

(title)

_____   _____   _____   _____   _____

- -

_____

(title)

_____   _____   _____   _____   _____

_____

(title)

_____   _____   _____   _____   _____

©Teacher Created Resources, Inc.

What's My Scale?

Skills

- Use small grid paper to construct column graphs
- Analyze column graph data to form opinions
- Use different scales on column graphs

Grouping

- whole class
- small groups

Materials

- small grids (page 32 or 33)
- graph starter cards (pages 35–37)
- graph discussion cards (page 38)
- scrap paper
- colored pencils

Directions

- Have students imagine they are collecting information about whether they like swimming in the sea or in a swimming pool. Have them imagine they are able to get information from 100 people. Ask, "How can you graph large numbers like this? What will you do if you have 46 people say they like swimming in the sea best?" (e.g., Use a grid with really tiny spaces.)
- Tell students one way to graph really large numbers is to use a different scale. (e.g., Instead of saying each tally mark equals one space, they can say it equals two. Every time they get two responses, they can color in one space. If they get three responses, they can color in one space and half of the next space.) Ask, "How many responses will two spaces represent?" (2 x 2 = 4) Ask, "How many responses will 10 spaces represent?" (10 x 2 = 20) Or, tell students they can call each space 5. That way they can fit a lot of numbers into a smaller graph. (e.g., To show 27 responses on a scale like this, color five spaces and a little bit more of the next space to show the extra two responses.) Or, tell students they can call each space 10 so they can easily show up to 100 responses in each column.
- Have students work in small groups. Have them decide on a graph question to explore. Have students find a way to collect and record their information over a week or more. (e.g., Perhaps they can ask five people each day.) Have students decide what type of grid they will need to present their findings as a graph.

Variations

- Have students complete a graph using one of the graph starter cards (pages 35–37) with their teams.
- Have students look at one of the graph discussion cards (page 38) with their team. Ask, "What would be a suitable title? What do the labels represent?" Have students analyze and discuss the results together.

34

©Teacher Created Resources, Inc.

Parents

My parents were born in Malta.

Where were your parents born?

Seasons

I love summer best because that is when I can swim almost every day.

What is your favorite season?

My Worst Nightmare

What is the scariest thing you can imagine?

Ask everyone to reveal their worst fears.

Chocolate-Covered Candies

There are many colors in a bag of chocolate-covered candies.

Which one do you think is the most common?

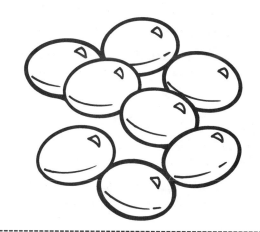

©*Teacher Created Resources, Inc.*   *#3533 Math in Action*

Fold Your Arms

When I fold my arms, I put my left arm over my right.

Which way do you fold?

Bed Time

I go to bed at 8:30 P.M.

What is your bedtime?

Bands

Many years ago the Beatles was the most popular band in America.

Which is your favorite band?

Dinosaurs

I think Iguanadon is fascinating.

Which dinosaur do you find most interesting?

#3533 Math in Action   ©*Teacher Created Resources, Inc.*

Traffic

What sort of vehicles pass your school gate over the next 30 minutes?

Running Steps

How many steps do you take when you run from here to the end of the playground over there?

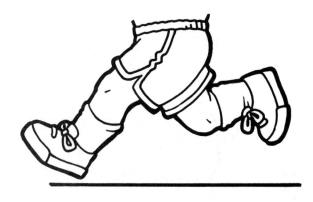

Computer Games

What are the top five favorite games played in your class?

Melt It

What takes longer to melt—ice, chocolate, butter, or ice cream?

©*Teacher Created Resources, Inc.*

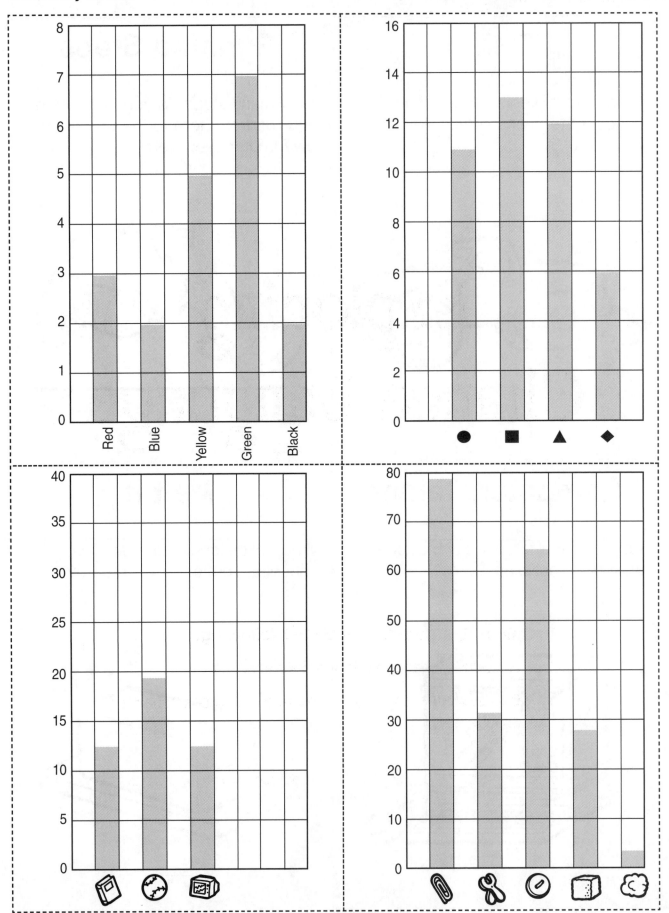

Exploring Chance Language

In this unit, students will do the following:

- Recognize chance events in daily activities
- Use everyday language to describe and predict chance events

(The skills in this section are listed on the Skills Record Sheet on page 94.)

What's the Chance?

Skills

- Recognize chance events in daily activities
- Use everyday language to describe and predict chance events

Grouping

- whole class
- small groups
- pairs

Materials

- What's the Chance? discussion cards (page 41)
- What's the Chance? labels (page 42)

Directions

- Ask students, "How do you come to school each day?" (e.g., walk, drive, catch a bus, ride a bike)
- Ask students, "Do you always arrive this way or are there other ways to arrive? What is the chance that you arrive by taxi? Come by a rowboat? Arrive by rollerskates? Parachute out of an airplane? Why? Why not?" Discuss with students using expressions like "There's a good chance" or "There's a poor chance."
- Ask, "What are other unlikely ways you might arrive at school?" Have students list these and explain to a partner why each event has a poor chance of happening. Ask, "Are any events impossible?"
- Have students play What's the Chance? with a partner. Have them ask their partners questions about chance events. (e.g., "What's the chance you will eat a banana for lunch? What's the chance you will wear something red tomorrow?") Their partners can only answer, "There's a good chance," "There's a poor chance" or "There's no chance."

Variations

- Have students work in small groups. Have them shuffle the What's the Chance? cards and place them face down in the center. Student take turns turning over the top card and telling their partners the chance of arriving at school that way.
- Using the three chance labels, have students sort the 12 cards into three groups. Have them explain to a friend why they sorted the cards like this.
- Using the three chance labels, have students list or draw their own events for another group to sort.

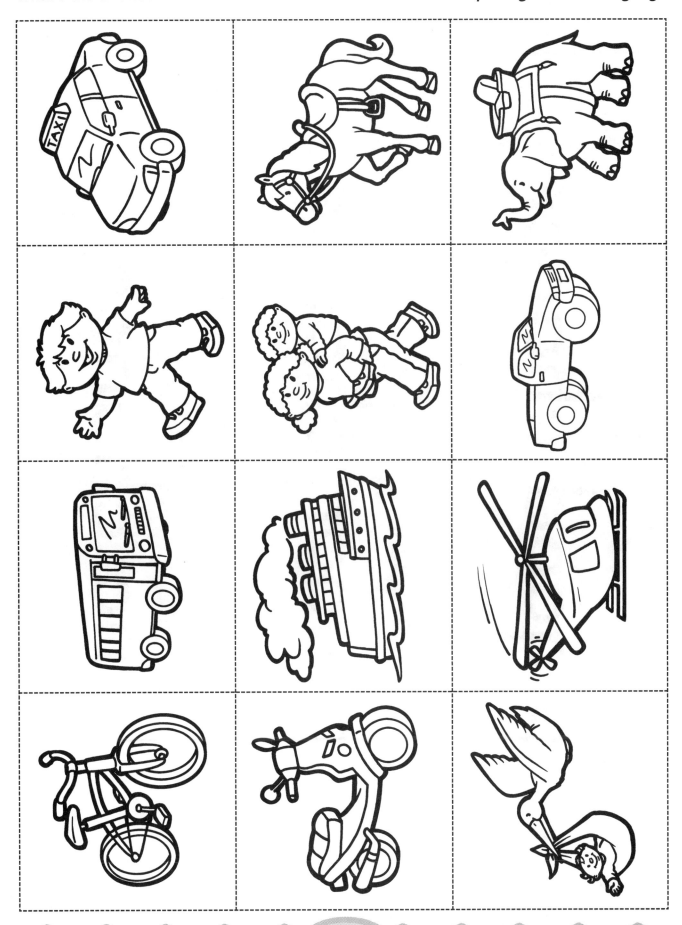

©*Teacher Created Resources, Inc.*   *#3533 Math in Action*

a good chance

a poor chance

no chance

#3533 Math in Action   *©Teacher Created Resources, Inc.*

Never Ever?

Skills

- Recognize chance events in daily activities
- Use everyday language to describe and predict chance events

Grouping

- whole class
- small groups
- pairs

Materials

- Never Ever? labels (page 44)
- workbooks
- pencils

Directions

- Ask students to close their eyes. Have them think of something that never happens. (e.g., I will grow three legs. A baby will be my teacher tomorrow. Our family will win the lottery tonight.) Discuss each suggestion together. Ask, "Why do you think this never happens? Could this ever happen? Is it impossible? Or has it just not happened yet?"

- Tell students to think of something that always happens. (e.g., The sun will rise in the morning. I will be one day older tomorrow.) Discuss each suggestion together. Ask, "Could this ever not happen? Why? Why not?"

- Explain to students that not everything in life happens so regularly. Some things happen often, but not always. (e.g., My dog jumps up on me when I arrive home.) Discuss with students at least five more events like this.

- Explain to students that some events happen irregularly. Sometimes they do happen, sometimes they do not happen. (e.g., It will rain tomorrow.) Discuss with students at least five more events like this.

- Explain to students that some events happen so irregularly that we say they are rare. (e.g., Grandpa will pick me up from school today.) Discuss with students at least five more events like this.

- Have students shuffle the five label cards. Have them select one at random (e.g., *never*), then think of an event to match. Have them each whisper their events to the persons sitting next to them. Try this at least two more times. Ask, "Can you think of your next story in less than a minute?"

- Have students draw pictures to show these chances. Have each student describe, to a partner, what is happening. Then have them write the labels to match.

Variation

- Have students make a class display. Have them draw events that will never happen, sometimes happen, or always happen.

never

rarely

sometimes

often

always

Is It Possible?

Skills
- Recognize chance events in daily activities
- Use everyday language to describe and predict chance events

Grouping
- whole class
- small groups
- pairs

Materials
- large color pictures (posters of detailed indoor or outdoor scenes)
- picture card (page 46)
- Is It Possible? discussion cards (page 47)

Directions
- Have students think of tonight's dinner. Ask, "What are some possible meals to eat?" (e.g., stir-fried vegetables, chicken and rice, a salad and garlic bread) Discuss with students at least five ideas.
- Ask, "What are some impossible meals to eat?" (e.g., "I really do not like liver, so I will never eat that!" or "I will never eat rocks for dinner.") Discuss with students at least five ideas.
- Tell students that there are many things in life that we have not yet done, but they still remain possible. (e.g., "I could climb to the top of Mount Everest, but it is very unlikely.") Have students think of more events like this that are not likely but still possible to achieve.
- Tell students that there are some things in life that are definitely impossible. They can never happen—ever! (e.g., a donkey flying with wings) Have students think of more events like this.
- Investigate with students the activities shown on a poster. Tell them to look at what is happening now. Ask, "What could happen next? Look for events that could possibly happen. Then look for events that could never happen." Tell students to use questions like "What if . . .?"

Variations
- Use the picture card (page 46) if you do not have a suitable color poster. Ask students to think of questions to ask a friend relating to possible and impossible. (e.g., "Could the cat catch the mouse?")
- Have students look at the 12 discussion cards. Ask them, "Do they show possible or impossible events? Can you think of a way that each can become possible?" (e.g., The flying girl might be using a trampoline or a diving board.)
- Have students sort the 12 discussion cards using the Never Ever? labels (page 44).

Weather Watch

Skills

- Recognize chance events in daily activities
- Use everyday language to describe and predict chance events

Grouping

- whole class
- small groups
- pairs

Materials

- Weather Watch cards (page 49) enlarged, colored, and laminated
- Weather Watch chart (page 50)
- workbooks
- colored pencils

Directions

- Have students go outside and look at the sky. Ask, "What are all the weather clues you can discover? Does the weather look the same all over?"
- Ask, "What do you think the weather will be like tomorrow?" Have students predict first, then explain their predictions.
- Have students look at the six Weather Watch cards (page 49). Tell students to think of tomorrow's weather. Have them put them in order from the likeliest to the least likely. Have them use chance words like *a good chance*, *possible*, *impossible*, *sometimes*, *rarely* to explain their choices. Have students check these again at the end of tomorrow's lessons.
- Have students draw a picture depicting what they think the weather will look like tomorrow. Discuss with them how this may change the way they will dress, what they will eat, and what they will do. Ask, "How else might the weather affect us tomorrow?"

Variations

- Have students investigate weather predictions in the local newspaper, radio, or TV reports.
- Have students use the Weather Watch chart (page 50) to keep a record of the daily weather over four weeks. At the end of this time, have them construct a graph to show their results. Tell students to think of three statements to make about the weather, based on this graph. Encourage them to use "chance words," as in direction above.

©Teacher Created Resources, Inc.

	Week 1	Week 2	Week 3	Week 4
Monday				
Tuesday				
Wednesday				
Thursday				
Friday				
Saturday				
Sunday				

That's Not Fair

Skills

- Recognize chance events in daily activities
- Use everyday language to describe and predict chance events

Skills

- whole class
- small groups
- pairs

Materials

- a chair
- counters (see directions)
- plastic farm and zoo animals, dinosaurs
- That's Not Fair game board (page 52)
- a bag of stickers
- die
- paper bags

Directions

- Measure the heights of two students. One student stands on the floor, the other stands on a chair. Ask students, "Who is the tallest? Is this fair? Why? Why not?"
- Share some stickers among three students. Give most to one and a few to the others. Ask students, "Is this fair? Why? Why not?"
- Play a game with the students as follows: Put six blue counters and two red counters in a bag. You get one point if you select blue. The class gets one point if you select red. Select a counter from the bag. Each time you select a counter, tally the score on the board and return the counter to the bag. Ask, "Who has the most points at the end of 10 turns?"
- Ask students, "Is this game fair? If not, what is unfair about it? Can you make it even more unfair?" (e.g., You get two turns before the students get their turn. You get five points for red.)
- Ask students, "How can you make it fair?" (e.g., Put the same number of counters in the bag. Each person gets one turn at a time.)
- Have students play a game with a partner. They will need a paper bag with two sets of five interesting counters. (e.g., five farm and five zoo animals) Have them design a game for two that is fair. Have them design another game that is unfair. Have students play one of their games for about five minutes. Have students discuss the results with another team.

Variations

- Design some outdoor races. Make some fair and some unfair. Discuss the results after playing for a set time limit. (e.g., a 20-yard race where some students start halfway)
- Have students play That's Not Fair (page 52) with 2–4 players as follows: Throw the die. If you land on ★, move forward. If you land on ●, move back. Try to be the first to finish. Ask students, "How can you vary the rules to make it really unfair?"

That's
Not
Fair

START | END

#3533 Math in Action   ©*Teacher Created Resources, Inc.*

That's Lucky

Skills

- Recognize chance events in daily activities
- Use everyday language to describe and predict chance events

Grouping

- whole class
- small groups
- pairs

Materials

- a set of name cards for each student (e.g., see page 14)
- a pack of cards
- four counters
- Good Luck game board (page 54)

Directions

- Ask students, "What could you do to make someone in your class really happy?" (e.g., let him or her ring the bell for lunch, tell him or her how much you like him or her, give him or her an award) Discuss suggestions. Have students select one that is possible to do. Put all of the name cards in a bag. Draw out one name. This student gets the award.

- Ask students, "Was this a fair contest? Did you have a good, a poor, or no chance of winning? If there is an equal chance of someone winning, you say this person is lucky and all the others are unlucky."

- Ask students, "What are some other lucky events?" (e.g., winning the lottery) Ask, "Is it lucky if your big sister wins a singing competition? Why? Why not?"

- Have students investigate different ways to go first in an activity. (e.g., turn over a red card in a pack, select the hand which has a hidden counter, say a counting-out rhyme, throw the highest number on a die) Try some of these for the rest of the day's tasks.

Variations

- Have students investigate the history of lucky charms. (e.g., a rabbit's foot, a four-leaf clover, a good luck wish, a horseshoe)

- Have students play Good Luck. They will need two to four players, four counters, a pack of playing cards and the Good Luck game board (page 54). The directions are as follows: Start in the center and each select a clown. If a player turns over a red card, he or she moves forward one space. If he or she gets black, he or she moves back one space. Tell them to each try to be the first to reach their clowns. Ask, "Were you lucky or unlucky? Was this a fair game? Can you play this again in an unfair way?"

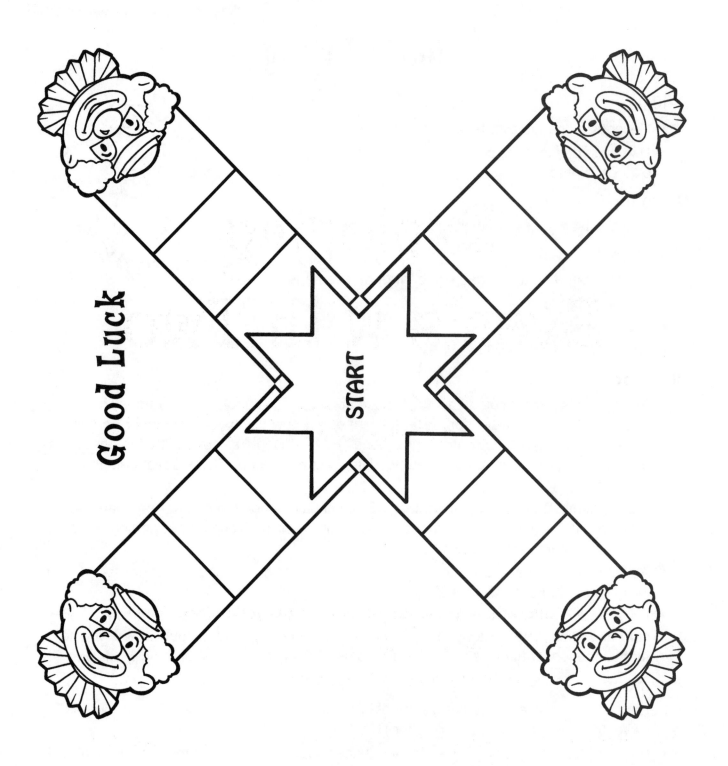

Good Luck

START

Exploring Events with Two Outcomes

In this unit, your students will do the following:

- Identify events with two outcomes

- Design a way to test a chance event

(The skills in this section are listed on the Skills Record Sheet on page 94.)

Is It Likely?

Skills
- Identify events with two outcomes
- Use everyday language to describe and predict chance events

Grouping
- whole class
- small groups

Materials
- workbooks
- pencils
- Is It Likely? discussion cards (page 57)

Directions
- Have students think of three things they will do when they go home today. (e.g. change their clothes, eat some fruit, talk to their moms) Ask students, "How do you know for sure that you will do these things? Is there any chance that you won't?" (e.g., their shorts might be in the wash, they might have run out of fruit, mom might be staying late at her work)

- Have students think of three things they are unlikely to do this afternoon. (e.g., visit a circus, wash the car, visit Grandpa) Ask, "Why are these things unlikely?"

- Tell students that when they think of something in the future, they think they know whether it is likely or unlikely to happen. Have them each face a partner and think of a question to ask him or her. (e.g., "Will you see a movie tonight?") They answer either "That is likely to happen" or "That is unlikely to happen." Then discuss their reasons. Ask, "How many questions like this can you answer in a set time limit?" (e.g., three minutes)

- Have students draw two different pictures—something that is likely to happen and something that is unlikely to happen. Have them write explanations for each picture.

Variations
- Have students play Is It Likely? (page 57) in small groups. Have them shuffle the discussion cards and scatter them face down in the center. Students take turns turning over any card and telling their partner whether the event shown is likely or unlikely to happen. Have students give reasons for their statements.

- Have students play This Is More Likely. Have them turn over any two cards. Have them look at the two events and say which one is more likely to happen. Have them make up some more cards like this to add to the pile.

- Have students sort three or more cards into order from the most likely to the least likely.

I will eat potatoes for my dinner.

It will be quite sunny tomorrow.

I will be two inches tall when I am 20.

There will be a visitor at our place this weekend.

I will see a friend on the way home from school.

I will be an astronaut when I grow up.

My cat will have three kittens.

The principal will come into our room soon.

I will keep my bedroom tidy.

I will win the next game I play.

What's a Fair Test?

Skills

- Identify events with two outcomes
- Design a way to test a chance event

Grouping

- whole class
- small groups

Materials

- What's a Fair Test? discussion cards (page 59)

Directions

- Tell students that when someone has a baby, it will be either a boy or a girl. There are only two possibilities (unless someone has twins or triplets). Tell them that some people think the same numbers of boys and girls are born each year. Ask, "What do you think? How can you find out?" (e.g., Check census statistics on the Internet.)

- Tell students that one way is to design a test to find out which babies have been born to parents in our school this year. Ask students, "Whom should you ask? How will you collect the information? How will you present what you find?" Discuss a variety of suggestions.

- Ask students, "What makes a fair test? If you only ask two people, is this fair? If you ask 10 people, is this fair?" Discuss together. Have students decide on the smallest number of people they will ask. (e.g., ask at least 20 people) Have students decide how they will record their results. (e.g., draw a grid and make tally marks)

Boys	
Girls	

- Ask students, "What are some other ideas to test that have only two possible outcomes?" (e.g., Sneakers are the most popular shoe. Dogs are the most popular pet.) These statements have only two possibilities. (e.g., They are either true or not true.) In teams, have students decide on a statement that has only two outcomes. Have them design a fair test for this. Have them think about how they will carry out their test and how they will report their results.

Variation

- Have students design a fair test for one of the statements on the 10 discussion cards (page 59).

When you smile at someone, he or she smiles back.

When you say "hello" to someone, he or she says "hello" back.

A car will pass by in the next two minutes.

People start walking with their right feet.

If you fold your arms when you are talking to someone, he or she will fold his or her arms, too.

A buttered slice of bread will fall face down.

Everyone loves to eat broccoli.

The next person to come around the corner will be a girl.

Chocolate is the favorite ice-cream flavor.

People enjoy reading books more than watching TV.

©*Teacher Created Resources, Inc.*

Cookies

Skills

- Identify events with two outcomes
- Design a way to test a chance event

Grouping

- pairs

Materials

- a Cookies worksheet (page 61) for each player
- two cardboard copies of the Cookies playing cards (page 62)
- paper
- pencils

Directions

- Ask students, "What is your favorite cookie? Do you prefer homemade cookies or cookies bought in a store? What type of cookie do you think will be the favorite in your class?"

- Have students play the Cookies game with a partner. Have them look at the two types of cookie cards. Have them decide on a flavor for each one. (e.g., chocolate chip and orange fudge) Have each student choose one type of cookie. Then have one person shuffle the cards and place the cards face down in the center. Students take turns taking the top card. If it matches his or her cookie, he or she places it in his or her cookie jar. After 10 cards have been selected, have students count how many cookies are in their jars. Ask, "Did you each get a fair share?"

- Ask students, "How can you change the rules so that this game is no longer fair?" (e.g., Use 12 round and six square cookie cards.)

- Have students design their own games using two types of food. Have them make game boards and cards and ask some friends to test the games.

Variations

- To play Cookies in pairs with up to the whole class, give each pair a worksheet and one round and one square cookie card in a paper bag. When it is a student's turn, he or she pulls one card from the bag. The person who has a match for this card draws a cookie on his or her worksheet. Replace the card and continue for a set number of turns. (e.g., until someone has drawn five cookies on their sheet.)

- Have students find recipes for their two favorite cookies. Have them make batches of each to sell during recess or lunch or at the school cafeteria. Have students predict which type will be the favorite, then have them test their prediction by collecting their sales information. Have them think of two statements to make about their results.

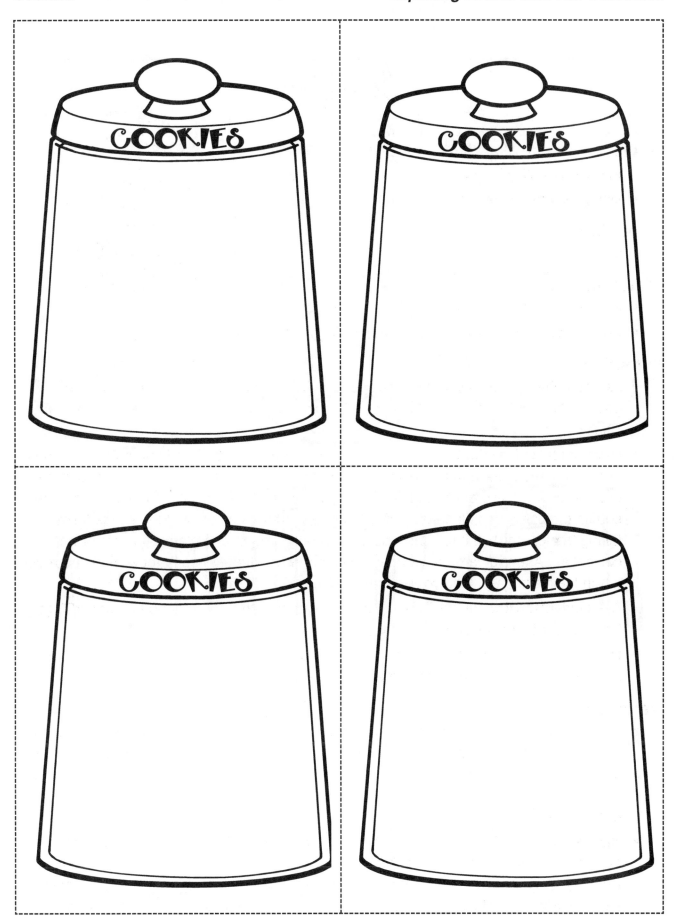

©*Teacher Created Resources, Inc.*

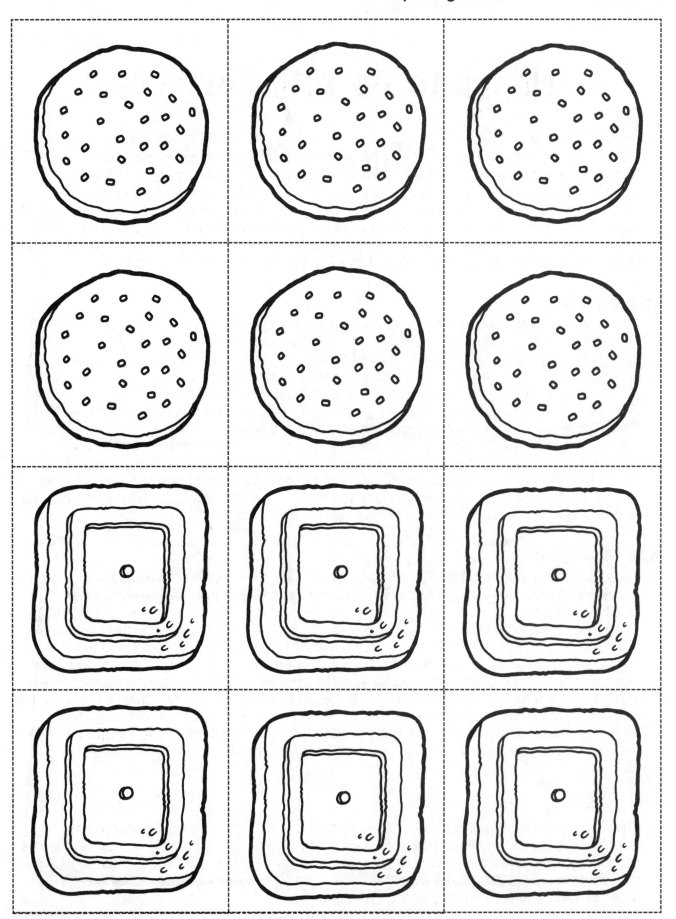

Things to Do with Spinners

Bottle Spinners

(*Whole Class*) Have students sit in a large circle. Form two teams by dividing the circle in half (draw a line down the center of the circle with chalk or masking tape). Select someone to spin an empty, plastic two-liter soda bottle in the center of the circle. Ask students, "What chance do you have of the bottle pointing to your half of the circle when it stops?" Tell them that their team scores five points if the bottle points to their team's half of the circle when it stops. Tally the points, and at the end of ten spins, count to see which team has the most points. Ask, "Was this a fair outcome? If not, why not?"

Spin My Web

(*Two Players*) Reproduce the Two-Color Spinners (page 64) on heavy cardstock. Give one spinner and two Spin My Web cards (page 65) to each pair of students. Have each student choose a side of the spinner and color it their favorite color (the two sides should be different colors). Then, show students how to use a paper clip and a pencil to create a spinner. (Place the end loop of the paper clip over the tip of the pencil, hold the tip of the pencil to the dot in the center of the spinner, and flick the paper clip with a finger to spin it around the pencil.) Ask students, "What chance do you have of the spinner landing on your color?" Each student takes a turn to spin the spinner. If the paper clip lands on the player's favorite color, he or she can color in one section of the web on his or her Spin My Web card. The first player to color in all sections of the web wins. Ask, "Was this a fair outcome? If not, why not?"

How Many Teeth?

(*Two Players*) Reproduce the Creature Spinners (page 66) on heavy cardstock. Give one spinner and one copy of How Many Teeth? creature cards (page 67) to each pair of students. Have each student choose whether he or she will be a crocodile or a shark. Then, show students how to use a paper clip and a pencil to create a spinner. (Place the end loop of the paper clip over the tip of the pencil, hold the tip of the pencil to the dot in the center of the spinner, and flick the paper clip with a finger to spin it around the pencil.) Ask students, "What chance do you have of the spinner landing on your creature?" Each student takes a turn to spin the spinner. If the paper clip lands on the player's creature, he or she can draw two teeth on their creature card. At the end of a time limit (e.g., three minutes), tell students to count by twos to see how many teeth they have altogether. The player with the most teeth on his or her creature is the winner. Ask, "Was this a fair outcome? If not, why not?"

Who Starts First?

(*Two Players*) Have each pair of students prepare a two-color spinner (page 64) as directed under Spin My Web (above). Have students use the spinner to decide who starts first to say their times tables facts for a given number or who starts first in a game of tic-tac-toe or who first tries to roll a hoop in a straight line for ten yards or more. Ask, "What other games or events could you start this way?"

Two-Color Spinner

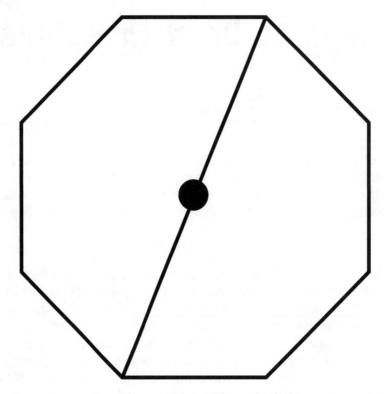

Two-Color Spinner

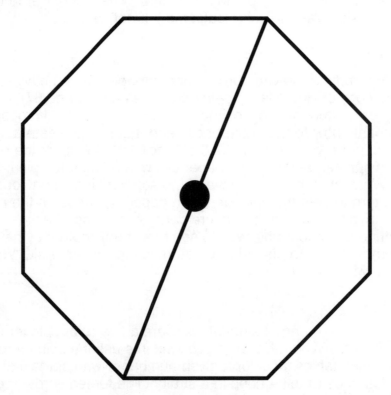

Spin My Web

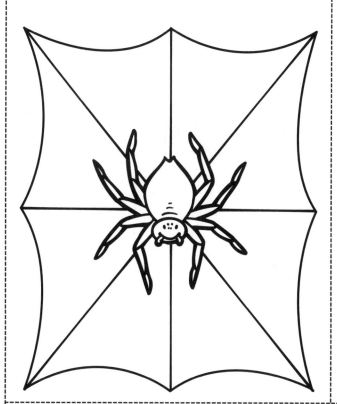

Spin My Web

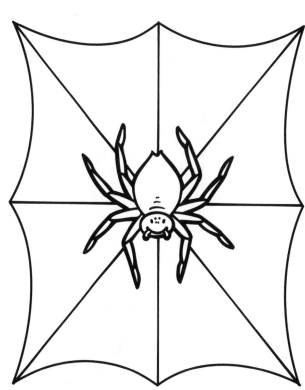

Spin My Web

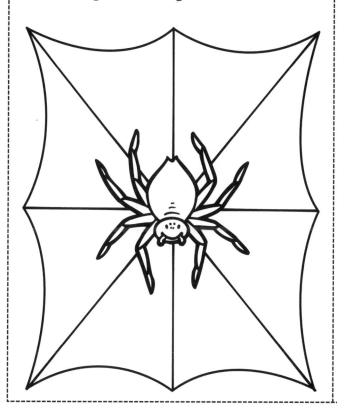

Spin My Web

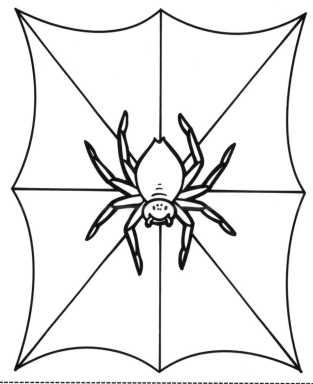

©Teacher Created Resources, Inc.

Creature Spinner

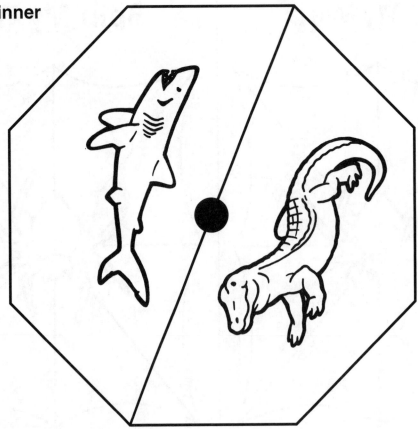

Creature Spinner

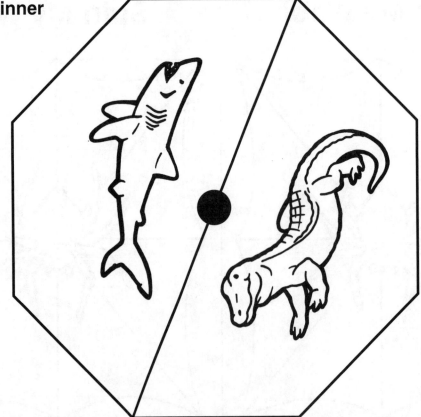

　　　　　©*Teacher Created Resources, Inc.*

How Many Teeth?

How Many Teeth?

How Many Teeth?

How Many Teeth?

©*Teacher Created Resources, Inc.*

Things to Do with Coins

How Far Away?

- *(Two Players)* Give each pair of students a yardstick or one yard of string, a starting point, and a coin. Tell them they are trying to move the greatest distance from the starting point. Ask "If you toss a coin, what is the chance that it will come up heads? Tails?" Have students decide whether they are heads or tails. Have them each toss the coin. If it lands on their side, they step forward one yard. After 10 tosses, how far is each student from the start? Have students estimate first, then check. Ask, "Who is the furthest away? Is this fair?" Tell students to invent their own rules, too. (e.g., Step forward if you get heads. Step backward if you get tails. Step five inches ahead each time you win the toss.)

Toss and Take

- *(Two Players)* Give each pair of students five coins. Tell them they are trying to collect as many coins as possible. Have them decide whether they are heads or tails. Say, "When it is your turn, toss one of your coins. What chance do you have of it landing your way? If you win the toss, you take one of your partner's coins. If you lose the toss, you give one of your coins to your partner. Who has the most coins at the end of 10 tosses? Is this fair?" Tell students to invent their own rules, too. (e.g., If you win, take two coins.)

Monkey Business

- *(Two Players)* Give each pair of students a game board (page 69), two small counters (these are the monkeys), and a coin. Tell them they are racing to reach the coconuts at the top of their palm trees. Have students decide whether they are heads or tails. Say, "When it is your turn, toss the coin. What chance do you have of it landing your way? If you win the toss, you climb one space up your palm tree. If you lose the toss, your partner climbs up one space on his or her palm tree. Who is first to reach the top? Is this fair?" Tell students to invent their own rules, too. (e.g., If you win the toss, climb up two spaces and make your partner climb down one space.)

What a Face

- *(Two Players)* Give each pair of students a worksheet (page 70). Tell them they are trying to draw in nine features on their face—hair, two ears, two eyes, two eyebrows, a mouth, and a nose—in any order. Have them decide whether they are heads or tails. Say, "When it is your turn, toss the coin. What chance does your partner have of it landing their way? Whoever wins the toss draws one feature on their face. Who is first to draw a complete face? Is this fair?" Tell students to invent their own rules, too. (e.g., Draw extras such as cheeks, a chin, and anything else you can think of.)

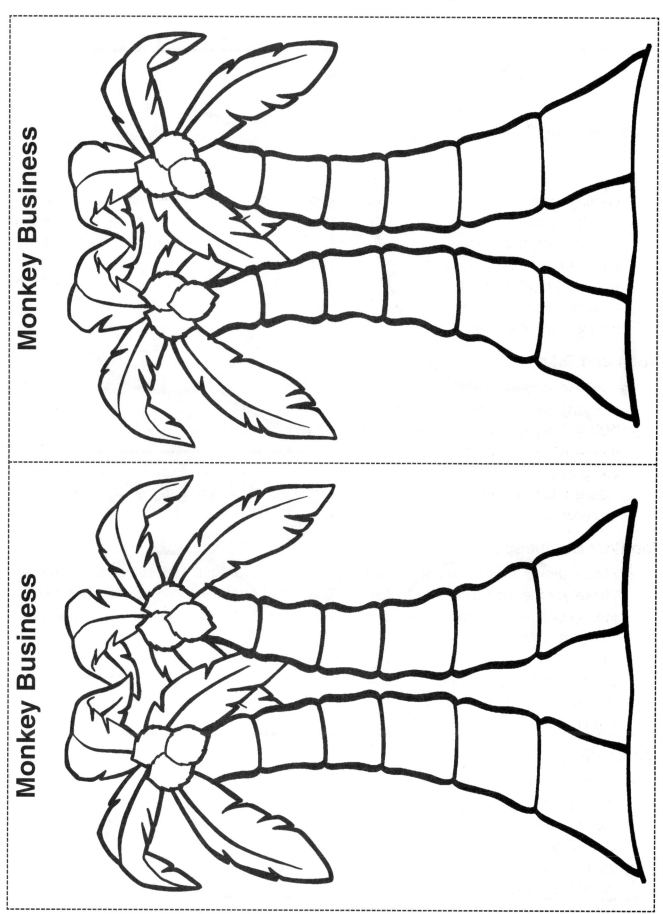

Monkey Business

Monkey Business

©Teacher Created Resources, Inc.

What a Face

What a Face

What a Face

What a Face

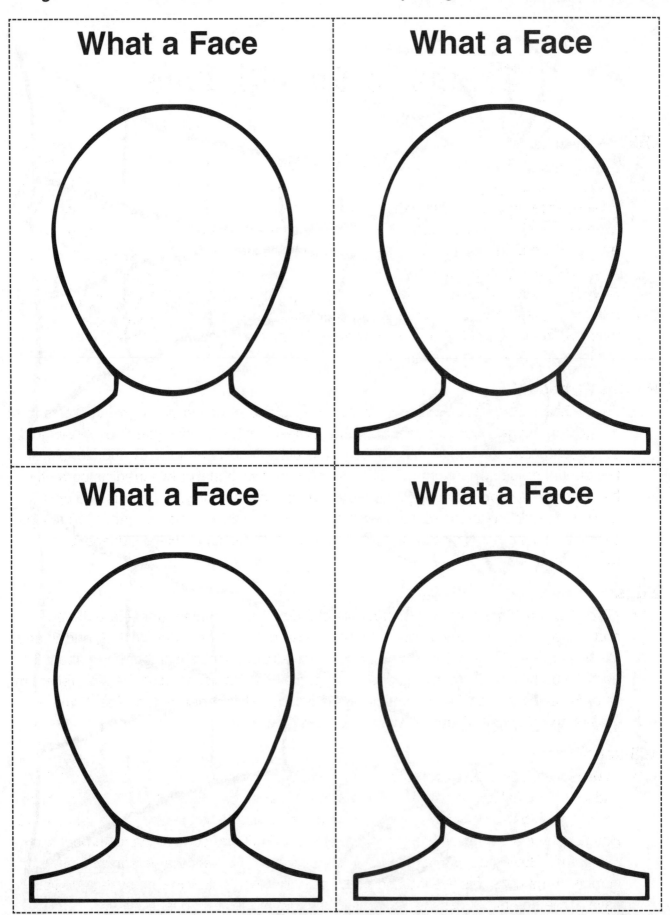

Things to Do with Dice

Odd or Even

- *(Two Players)* Give each pair of students one die, some scrap paper, and a pencil. Ask, "If you throw the die once, what chance do you have of getting an odd number? An even number?" Tell them to decide whether they are odds or evens. Have students throw the die. Tell them that the person whose type of number comes up scores 10 points. After 10 throws, tell them to add up their scores. Ask, "Who has the highest total? Is this fair?" Tell students to invent their own rules, too. (e.g., Throw two dice and add the numbers to get an odd or even total.)

Pirates' Treasure

- *(Two Players)* Give each pair of students one die and some yellow or gold counters for the golden nuggets from a treasure island. (These can be small squares of yellow/gold construction paper or bottle caps.) When it is player one's turn to throw the die, player two should predict what number will appear on the die. If player two predicts correctly, he or she wins two gold nuggets. If player two is incorrect, player one gets one gold nugget. Player one then gets to predict and player two throws the die. After five tosses each, ask, "Who has collected the most treasure? Is this fair?" Have students try tossing 10 times each.

Eggs

- *(Two Players)* Give each pair of students one die, some counters (these are the eggs), and two egg cartons. Tell students that the lower numbers on a die are 1, 2, and 3, and the higher numbers are 4, 5, and 6. Have each student decide whether he or she is going to be higher or lower throughout this game. Have students toss the die. Tell them whoever owns the matching number puts two eggs in his or her carton. Ask, "Who is the first to collect all 12 eggs? Were you lucky?"

Millipedes

- *(Two Players)* Give each pair of students one die, a Millipedes worksheet (page 72), and a pencil. Tell students that millipedes have a great number of legs. Tell them that the player with the most legs drawn on his or her millipede by the end of the game wins. Have each student decide whether he or she is going to be odds or evens throughout this game. Have students take turns tossing the die. Tell them the person whose number turns up draws five legs on his or her millipede. Ask, "How many legs do you have at the end of 10 tosses? Is it possible to get more than 30 legs?"

©*Teacher Created Resources, Inc.*

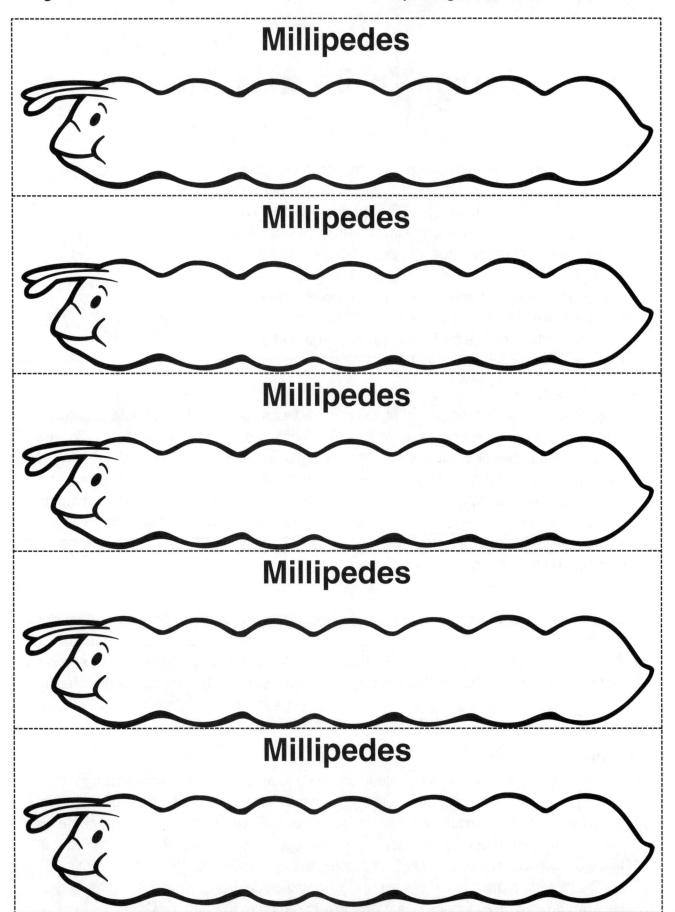

Millipedes

Millipedes

Millipedes

Millipedes

Millipedes

#3533 Math in Action   ©*Teacher Created Resources, Inc.*

Exploring Events with Three or More Outcomes

In this unit, students will do the following:

- Identify events with three or more outcomes
- List outcomes from chance experiments
- Order chance events

(The skills in this section are listed on the Skills Record Sheet on page 94.)

1 Chance in 3

Skills

- Identify events with three or more outcomes
- List outcomes from chance experiments

Grouping

- whole class
- small groups

Materials

- three containers
- small toy bear
- scrap paper
- paper bags
- Pets cards (page 75)
- pencils

Directions

- Talk with students about lucky and unlucky things that have happened to them lately.

- Secretly hide a bear under one of three containers. Have students guess where it is hidden. Ask, "What chance do you have of finding the bear's hiding place?" Discuss with the students the idea that there are three possibilities. They may be lucky and guess the correct one or they may be unlucky. Ask, "Is this fair?" (You have one chance in three of guessing correctly.)

- Have students play Which One Is It? in teams of three. Tell them to write their names on scraps of paper and fold them up. Have them put the three names in a paper bag. In turn, have students draw one name out. Ask, "What chance do you have that your name will be selected?" Have them return the papers to the bag and then draw out another piece. Repeat up to 10 times. Ask, "How often does your name appear? Is this fair?"

- Have students play Who Has It? in teams of four. Have them select a small object to hide. (e.g., an eraser or a die) One player is the guesser. The other three players put their hands behind their backs and hide the object in one player's hand while the guesser is not looking. When they call out "Who has it?", the guesser calls the name of the player he or she thinks has the object. Ask, "What chance does he or she have of guessing correctly? Is this fair?"

Variation

- Have students play Whose Pet? in teams of three. They will need the 15 Pets cards, shuffled, and then placed face down in the center. Have them choose to be either a cat, a dog, or a bird. Have them each turn over the top card 10 times. Tell them to keep tallies of how often their pets appear. Ask, "What chance do you have, each time, of the top card belonging to you?"

©*Teacher Created Resources, Inc.*

#3533 Math in Action

Dinosaur Danger

Skills

- Identify events with three or more outcomes
- List outcomes from chance experiments

Grouping

- small groups of three

Materials

- three counters
- Dinosaur Danger game board (page 78)
- Dinosaur Danger Spinner (page 77), pencil, paper clip, scissors

Directions

- Ask students, "What do you know about dinosaurs? Which was the fiercest dinosaur? The longest dinosaur? The fastest dinosaur? How long ago did they live on our planet?"
- Give each group a pencil, a paper clip, scissors, one spinner, three counters, and a game board.
- Have students cut out the spinners. Laminate for durability. Have them place a paper clip through a tip of a pencil. Then, place the pencil at the center of the spinner and check for bias. (See sample illustration.)
- Tell students to imagine they are each one of the three dinosaurs shown on the spinner. Ask, "What type of dinosaur are you? Do you eat meat or only plants?"
- Tell students to imagine that a huge volcano is about to erupt. Tell them they need to race back home to the safety of their cave.
- Have students take turns spinning the spinner. If the spinner lands on their dinosaur, they move from the volcano at "START" to the next space their dinosaur. If it does not match, they must wait where they are and try their luck next turn. Ask, "What chance do you have that the spinner will land on your picture?"
- Have students try to be the first dinosaur to reach the "FINISH" cave by spinning their matching pictures.
- Discuss the results of the game together. Ask, "Was it fair? Did one dinosaur appear more often than another?"

Variations

- Ask, "Does the same dinosaur always reach the cave first?" Have students play this game several times to check.
- Have students invent their own three-outcome game.

Dinosaur Danger Spinner

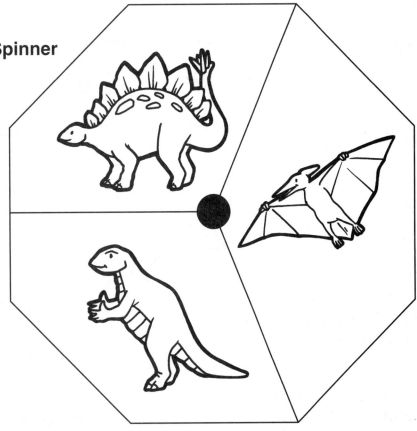

Dinosaur Danger Spinner

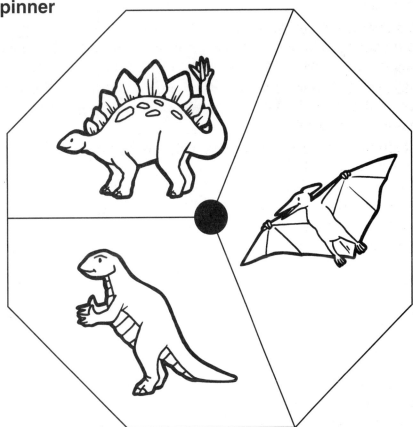

©Teacher Created Resources, Inc.   #3533 Math in Action

Hens

Skills

- Identify events with three or more outcomes
- List outcomes from chance experiments
- Order chance events

Grouping

- small groups
- pairs

Materials

- three different-colored marbles or beads
- Hens worksheet (page 80)
- paper bag
- colored pencils

Directions

- Tell students that when hens lay eggs, the eggs can be many different colors. (e.g., speckled, brown, white) The eggs can also be different sizes. (e.g., large, medium, or small)

- Tell students to each put three different-colored marbles in a bag. Have them imagine these are the possible egg colors. Have them color each of the three hens on the worksheets to match the colors they have selected.

- Have each student draw out one egg at a time, then replace it in the bag. Tell them to each keep a record of which color has been drawn by drawing an egg in the nest of the matching hen.

- Tell students to do this about 20 times, then count up how many eggs each hen has altogether. Ask, "Which hen laid the most eggs? Was this fair?"

- Have students put the hens in order by writing 1, 2, or 3 to show which hen laid the most. Tell them to discuss the results with their friends.

Variations

- Ask students, "What happens to your results if you put two eggs of each color in the bag (6 marbles altogether). Are the results what you expected?"

- Ask students, "How can you make this less fair? What happens if you put four of one color and one each of the other two colors in the bag?" Have students play Hens again and record their results.

Hens

Hens

Hens

Toss-a-Tangram

Skills

- Identify events with 3 or more outcomes
- List outcomes from chance experiments
- Order chance events

Grouping

- small groups of three

Materials

- two coins
- seven tangram pieces for each player (page 82)
- Toss-a-Tangram Game Board for each player (page 83)
- paper
- pencils

Directions

- Ask students, "What happens when you toss a coin?" (e.g., It can land heads up or tails up.) Ask, "What are the possible results if you toss two coins?" Discuss with students first, then have them check by tossing about 10 times and recording the results.

- Tell students they can use these results to figure out which of three players goes first in a game. For example, each player selects either HH (both heads), TH (tails and heads), or TT (both tails). Ask, "For what else could you use this chance event?" Discuss together.

- Tell students one chance game they can play is Toss-a-Tangram. Tell them each player needs a set of seven tangram pieces and a playing board. Have students select one of the three possible outcomes for a two coin toss (e.g., HH, TT, HT). This is their lucky symbol. Tell them to take turns tossing the two coins. Whoever has the matching lucky symbol on each toss places one piece of his or her tangram on his or her board. Students keep playing until one player completes his or her picture.

- Have students keep a record of the coin toss results for each turn. Ask, "Was this a fair outcome?"

- Have students place the three outcomes in order from the one that occurred the most to the one that occurred the least.

Variations

- Have each student create his or her own tangram playing board by rearranging the seven pieces to make a picture or a pattern. Have him or her trace the outline of these onto a sheet of paper.

- Have students make a faster game by creating a picture with fewer than seven pieces.

©*Teacher Created Resources, Inc.*

**Tangram
Pieces**

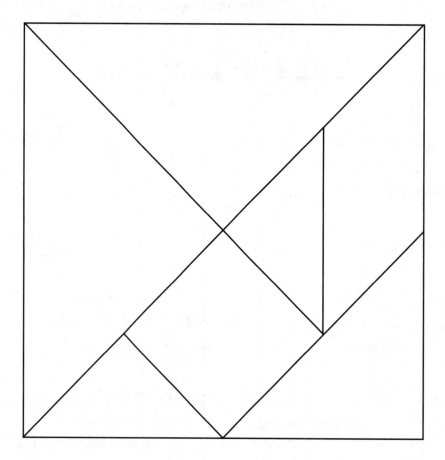

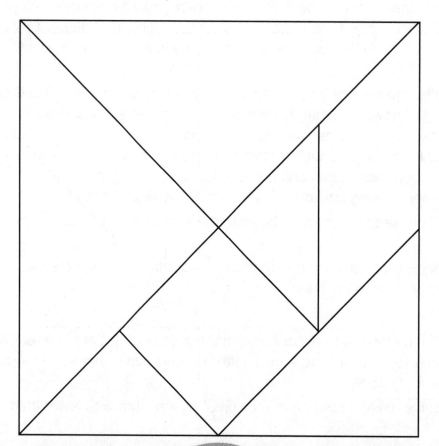

Toss-a-Tangram Game Board

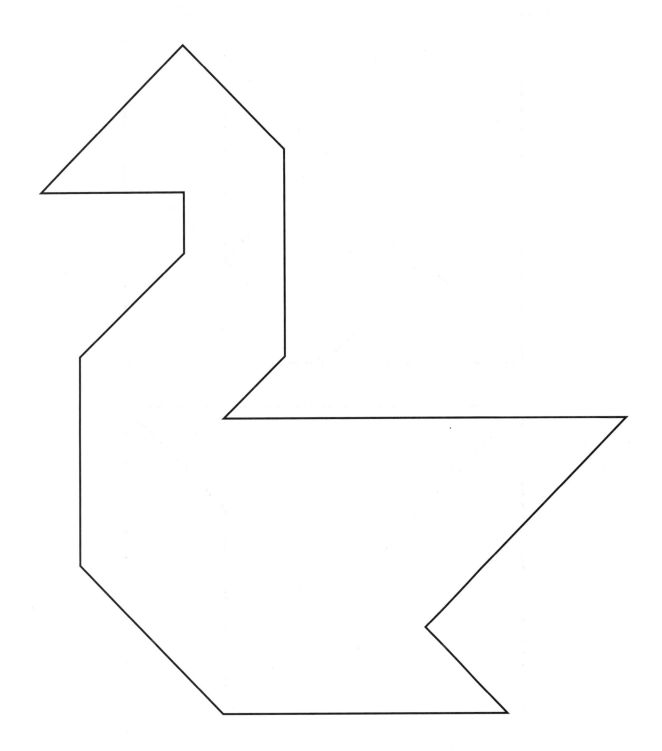

©*Teacher Created Resources, Inc.*

Which Suit Am I?

Skills

- Identify events with three or more outcomes
- List outcomes from chance experiments
- Order chance events

Grouping

- small groups

Materials

- a pack of playing cards
- Which Suit Am I? scoresheets (page 85)
- calculators
- paper
- pencils

Directions

- Have students tell you everything they know about a pack of cards. (e.g., What is a suit? How many different suits are there in a pack? What color are they? How many cards are in one suit? How many cards are in a whole pack?)

- Have a student shuffle a pack of cards. Ask, "What suit do you think the top card will be?" Have him or her turn over the top card 10 times and record the suit. Ask, "What do you notice? Is this fair? What chance do you have that a hearts card will appear?" (e.g., 1 chance in 4)

- Have student play Which Suit Am I? Have students shuffle a pack of cards and place it face down in the center. Tell them to take turns turning over the top card and tallying the suit on the Which Suit Am I? scoresheet (page 85). After each player has had five turns, have them add up their points. They score 10 points for each heart, five points for each club, two points for each diamond, and one point for each spade. Students can count their points by counting each tally by tens, fives, twos, or ones.

- Ask, "Who has the highest score? Is this fair? Did one suit appear more often than the others? Why do you think this happened?" (e.g., It's just luck. The pack wasn't shuffled very well.)

Variation

- Have students make up their own rules for Which Suit Am I? (e.g., If you turn over a joker, score 20 points.)

©*Teacher Created Resources, Inc.*

Which Suit Am I?

♥ 10	♣ 5	♦ 2	♠ 1
Total:	Total:	Total:	Total:

Which Suit Am I?

♥ 10	♣ 5	♦ 2	♠ 1
Total:	Total:	Total:	Total:

Which Suit Am I?

♥ 10	♣ 5	♦ 2	♠ 1
Total:	Total:	Total:	Total:

Toss-a-Turtle

Skills

- Identify events with three or more outcomes
- List outcomes from chance experiments
- Order chance events

Grouping

- small groups

Materials

- Toss-a-Turtle Game Board (page 87) enlarged, one per group
- four different counters
- Toss-a-Turtle Die (page 87) enlarged, cut out, folded, and pasted to form a cube OR draw a snake, turtle, fish, frog, and two blanks on a blank wooden cube—one per group
- paper or pencils

Directions

- Tell students games of chance are played all over the world. Ask them, "What are some chance games you know how to play?" (e.g., "Fish" or "Snap," dominoes)
- Tell students that on the island of Lombok in Indonesia, there is a game called "Epo." Tell them that each player places his or her token on one of four creatures carved into a wooden board—a snake, a turtle, a fish, or a frog. He or she then tosses a matching wooden die. If the die lands with his or her creature face up, he or she wins a point. If it lands on one of the two blank faces, no one wins a point. At the end of 10 tosses, players see who has won the most points.
- Tell the students that they can play their own version of "Epo" using the Toss-a-Turtle Game Board and Toss-a-Turtle Die. Have them tally their points on paper. Have them put the final results in order from the creature that appeared the most to the one that appeared the least.

Variation

- Have students investigate games of chance their parents or grandparents know how to play.

Toss-a-Turtle Die

Toss-a-Turtle Game Board

©*Teacher Created Resources, Inc.*

Find-a-Face

Skills

- Identify events with three or more outcomes
- List outcomes from chance experiments

Grouping

- small groups

Materials

- a pack of playing cards
- Find-a-Face cards (page 89)
- paper
- pencils

Directions

- Tell the students that in a pack of 52 playing cards (minus the jokers), there is a one-in-four chance that one will turn up a club card. Have students shuffle the cards. Have them turn over the top 12 cards. Ask, "Did clubs appear three times? Why? Why not?" (e.g., You need to turn over many more than 12 cards as even though one-in-four should be clubs, perhaps they are all at the back of the pack.) Tell students any card can turn up each time, but if you turned over all 52 cards, then one in four would be clubs.

- Have students play a version of this using the Find a Face cards. Ask, "What are the possible combinations when you turn over two cards?" (e.g., two left faces, two right faces, a left and right face that do not match, a left and right face that match)

- Tell students to take turns turning over any two cards. If they are two left or two right faces, score zero. If they are a left and a right face that do not match, score five points. If they are a matching left and right face, score 10 points. Have students keep a tally of their scores.

- Ask, "Who has the most points at the end of two turns each? Four turns each? Six turns each? Is this fair? What chance do you have of scoring the most points?" (e.g., just as good a chance as anyone else playing)

Variation

- Have students play a chance game using a pack of cards. Have students shuffle the cards and place them face down in the middle. Have students turn over the top card and place it face up beside the pack to start. When it is their turn, have them turn over the next card. If the suits of these two face up cards match, he or she wins all of the face-up cards and turns over a new top card for the next player. If the suits do not match, the next player turns over the top card and sees whether it matches. Ask, "Who has the most cards at the end of five turns each?"

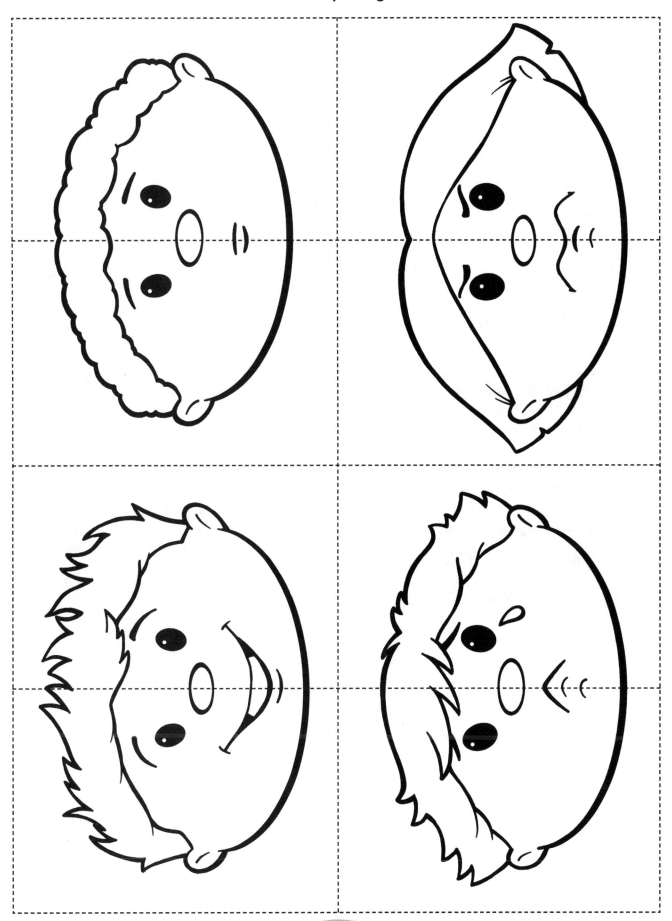

©*Teacher Created Resources, Inc.*

Sleeping Snake

Skills

- Identify events with three or more outcomes
- List outcomes from chance experiments

Grouping

- pairs

Materials

- Sleeping Snake Game Board (page 91)
- six tiny objects (e.g., paper clips, dried beans)
- a different game token for each player (e.g., counters, bottle caps)

Directions

- Tell the students that board games are played by most people in the world in some form or another. Tell them some are based on strategy, like Chess. Tell them that others are based on luck, like Snakes and Ladders, which is an old Indian game. Ask, "What board games do you know how to play?"

- Tell students Sleeping Snake is based on a board game played in Egypt over 4,000 years ago. Students are given six small objects (e.g., paper clips, dried beans). The "holder" divides the objects between his or her two clenched fists, and then shows them to the "guesser." The guesser points to a fist and guesses how many objects are inside. The holder shows the objects and together they determine the difference between the guess and the actual number of objects. The holder can then move his or her token along the snake board that number of spaces. (e.g., If the guesser guesses four, but the holder has only one, then the holder moves ahead three spaces because four minus one is three.) If the guesser guesses the number correctly, the holder does not move his or her token.

- The holder then gives the objects to the guesser and the next player in the circle becomes the guesser. Play passes around the circle. The first player to wake the sleeping snake by landing on its head wins.

- Ask, "Are the rules fair?" Have students play this game with some friends and discuss the results together.

Variations

- Have students make a faster game by hiding up to 10 objects between two fists.
- Tell students to vary the rules. (e.g., Use playing cards. If a card is diamonds, move four spaces; spades, move three spaces; hearts, move two spaces; or clubs, move one space.)
- Have students invent their own chance game and rules to match the game boards on pages 92 and 93. Ask, "What will you call each one?"

Sleeping Snake Game Board

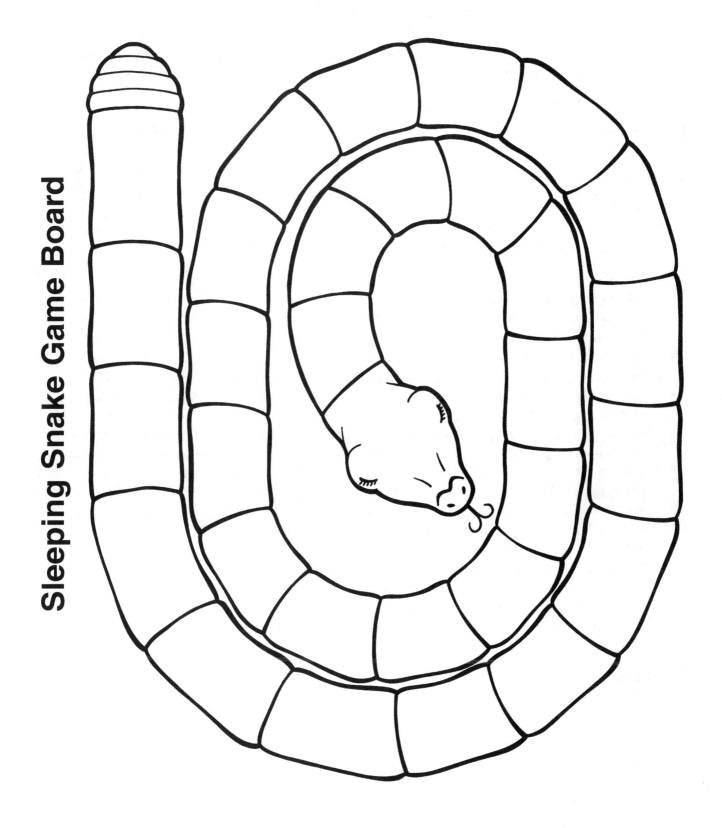

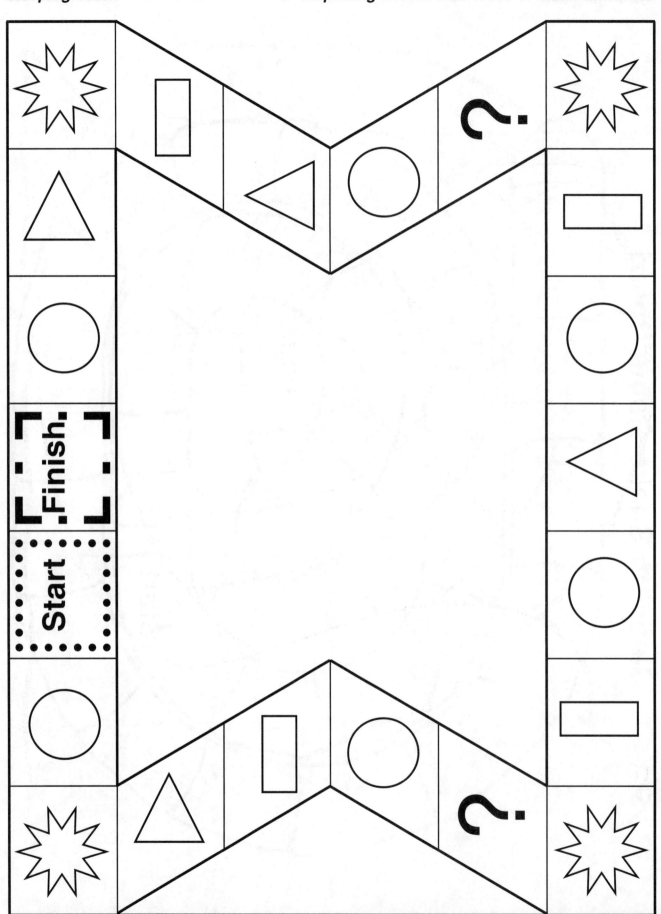

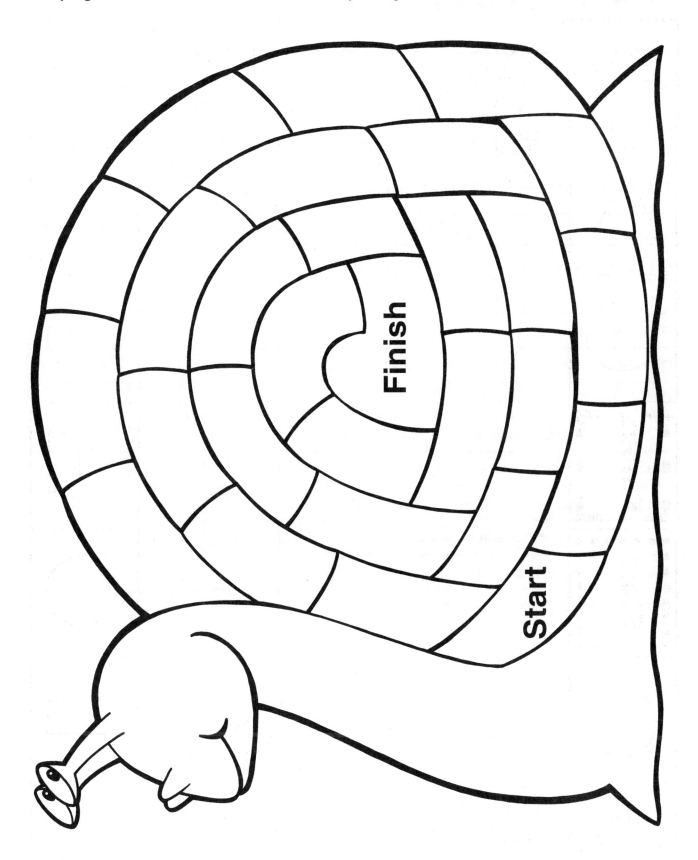

Finish

Start

Skills Record Sheet

EXPLORING GRAPHS, DATA AND CHANCE

NAME

Exploring Graphs								
Pose questions and collect related information								
Sort and compare groups by matching objects in lines								
Group pictures, symbols to represent, compare data								
Describe and interpret information from object displays								
Design and carry out a survey								
Place objects, pictures, and symbols in grids to represent data								
Use tally marks to collect data								
Use grids to construct, label, interpret column graphs								
Use small grid paper to construct column graphs								
Analyze column graph data to form opinions								
Use different scales on column graphs								
Exploring Data and Chance								
Recognize chance events in daily activities								
Use everyday language to describe and predict chance events								
Identify events with two outcomes								
Design a way to test a chance event								
Identify events with three or more outcomes								
List outcomes from chance experiments								
Order chance events								

94

©*Teacher Created Resources, Inc.*

Sample Weekly Program

STRAND Chance and Data

GRADE 1

SUBSTRAND Events with two outcomes

TERM 4 WEEK 3

SKILLS
- Identify events with two outcomes
- Design a way to test a chance event

LANGUAGE
- "likely," "unlikely to happen"
- "definitely will happen", "won't happen"
- "What's the chance?", "no chance," "a good chance"
- "possibility," "possible," "impossible"
- "a fair test," "fair," "unfair"
- "outcome," "result"

RESOURCES
workbooks
pencils
Is It Likely? discussion cards (page 57)

What's a Fair Test? discussion cards (page 59)
Cookies game board and cards (page 61 and 62)
paper bags

masking tape line in classroom, two-liter bottle, Two-Color Spinners (page 64), Spin My Web worksheet (page 65) Creature Spinners (page 66), How Many Teeth? worksheet (page 67)

coins, one-yard string, Monkey Business game board (page 69), counters, What a Face worksheet (page 70), pencils

dice, scrap paper, pencils gold bottle tops, counters, empty egg cartons Millipedes worksheet (page 72), pencils

MONDAY	TUESDAY	WEDNESDAY	THURSDAY	FRIDAY
• Review chance language (lucky/unlucky, never/sometimes/often). • Discuss events that you are sure will/won't happen. See Is It Likely? (page 56) • Group activities: A: Draw two events (likely/unlikely to happen) B: Is It Likely? discussion cards C: This is more likely • Whole class: Group reports and discussion	• Whole class: Discuss What's a Fair Test? (see page 58). • Discuss events with just two outcomes. Refer to What's a Fair Test? discussion cards in small groups. Design a test to carry out over the rest of this week. • Discuss suggestions. (e.g., How many people will you ask?) • Whole class: Explain rules for Cookies. Play in pairs using one square/one round card in a paper bag per pair.	• Whole class: Discuss what makes an event fair/unfair. • Play Bottle Spinners (page 63). Discuss results together. When could you use a spinner? • Group activities (page 63): A: Spin My Web B: How Many Teeth? C: Who Starts First? • Whole class recording: Write a report on the fairness of your activity outcomes.	• Whole class: Demonstrate and discuss what happens when you toss a coin (heads or tails). Are the results fair? • Group activities (page 68): A: How Far Away? B: Toss and Take C: Monkey Business D: What a Face • Whole class: Discuss your favorite activity. Explain how chance effects the outcomes.	• Whole class: Review outcomes using spinners, coins. Discuss lucky/unlucky. Demonstrate, discuss outcomes when tossing a die (1, 2, 3, 4, 5, or 6) • Group activities (page 71): A: Odd or Even B: Pirates' Treasure C: Eggs D: Millipedes • Whole class: Discuss your favorite activity. Review effects of chance in your daily life.

©Teacher Created Resources, Inc. #3533 Math in Action

Weekly Program

STRAND _____ SUBSTRAND _____

GRADE _____ TERM _____ WEEK _____

LANGUAGE

SKILLS

RESOURCES

MONDAY	TUESDAY	WEDNESDAY	THURSDAY	FRIDAY

©Teacher Created Resources, Inc.